What Is
BIOLOGY?

Rebecca Woodbury, Ph.D., M.Ed.

Gravitas Publications Inc.

What Is
Biology?

Illustrations: Janet Moneymaker

What Is Biology?
ISBN 978-1-950415-48-9

Published by Gravitas Publications Inc.
Imprint: Real Science-4-Kids
www.gravitaspublications.com
www.realscience4kids.com

RS4K

Photo credits: Cover and title page: By kieferpix, AdobeStock; P.1. By Rawpixel.com, AdobeStock; P.3. By Hamilton, AdobeStock; P.5. RociH from Pixabay; P.9. By frank29052515, AdobeStock; P.11 WikiImages from Pixabay; P.17 natasha_positive0 from Pixabay; P.19. By ondrejprosicky, AdobeStock; P.21. By Rawpixel.com, AdobeStock

Have you ever watched

a caterpillar eat a leaf?

Have you ever planted a
bean and watched it grow?

Have you ever looked at an ant carrying away your pie?

Ants eat pie?

Caterpillars, plants, ants,
and kids are all **living things.**

Hey!
Mice are living
things too!

Biology is the

study of living things.

I would like to
do that too.

Scientists who study biology are called **biologists.**

Hello!

Spiders are amazing!

Biologists study how
plants make food.

I want a Swiss
cheese plant.

Biologists study how
fish swim in the ocean.

Sometimes
I rest.

Biologists study how
tigers live in the jungle.

I don't like water!.

Biology and biologists help
us learn about living things.

I wonder how birds can fly.

How to say science words

biology   (biy-AH-luh-jee)

biologist   (biy-AH-luh-jist)

caterpillar   (KAA-tuhr-pih-luhr)

science   (SIY-uhns)

scientist   (SIY-uhn-tist)

study   (STUH-dee)

www.ingramcontent.com/pod-product-compliance
Lightning Source LLC
Chambersburg PA
CBHW040153200326
41520CB00028B/7592